The Machine That Saved The World:
Special Edition
Murray Leinster

Copyright © 2016 epubBooks

All Rights Reserved.

This publication is protected by copyright. By payment of the required fees, you have been granted the non-exclusive, non-transferable right to access and read the text of this ebook on-screen or via personal text-to-speech computer systems. No part of this text may be reproduced, transmitted, downloaded, decompiled, reverse engineered, stored in or introduced into any information storage and retrieval system, in any form or by any means, whether electronic or mechanical, now known or hereinafter invented, without the express written permission of epubBooks.

www.epubbooks.com

The Machine That Saved the World

They were broadcasts from nowhere—sinister emanations flooding in from space—smashing any receiver that picked them up. What defense could Earth devise against science such as this?

Did the broadcasts foretell flesh–rending supersonic blasts?

The first broadcast came in 1972, while Mahon–modified machines were still strictly classified, and the world had heard only rumors about them. The first broadcast was picked up by a television ham in Osceola, Florida, who fumingly reported artificial interference on the amateur TV bands. He heard and taped it for ten minutes—so he said—before it blew out his receiver. When he replaced the broken element, the broadcast was gone.

But the Communications Commission looked at and listened to the tape and practically went through the ceiling. It stationed a monitor truck in Osceola for months, listening feverishly to nothing.

Then for a long while there were rumors of broadcasts which blew out receiving apparatus, but nothing definite. Weird patterns appeared on screens high-pitched or deep-bass notes sounded—and the receiver went out of operation. After the ham operator in Osceola, nobody else got more than a second or two of the weird interference before blowing his set during six very full months of CC agitation.

Then a TV station in Seattle abruptly broadcast interference superimposed on its regular network program. The screens of all sets tuned to that program suddenly showed exotic, curiously curved, meaningless patterns on top of a commercial spectacular broadcast. At the same time incredible chirping noises came from the speakers, alternating with deep-bass hootings, which spoiled the ju-ju music of the most expensive ju-ju band on the air. The interference ended only with a minor break-down in the transmitting station. It was the same sort of interference that the Communications Commission had thrown fits about in Washington. It threw further fits now.

* * * * *

A month later a vision–phone circuit between Chicago and Los Angeles was unusable for ten minutes. The same meaningless picture–pattern and the same preposterous noises came on and monopolized the line. It ceased when a repeater–tube went out and a parallel circuit took over. Again, frantic agitation displayed by high authority.

Then the interference began to appear more frequently, though still capriciously. Once a Presidential broadcast was confused by interference apparently originating in the White House, and again a three–way top–secret conference between the commanding officers of three military departments ceased when the unhuman–sounding noises and the scrambled picture pattern inserted itself into the closed–circuit discussion. The conference broke up amid consternation. For one reason, military circuits were supposed to be interference–proof. For another, it appeared that if interference could be spotted to this circuit or this receiver it was likely this circuit or that receiver could be tapped.

For a third reason, the broadcasts were dynamite. As received, they were badly scrambled, but they could be straightened out. Even the first one, from Osceola, was cleaned up and understood. Enough so to make top authority tear its hair and allow only fully–cleared scientific consultants in on the thing.

The content of the broadcasts was kept considerably more secret than the existence of Mahon units and what they could do. And Mahon units were brand–new, then, and being worked with only at one research installation in the United States.

The broadcasts were not so closely confined. The same wriggly patterns and alien noises were picked up in Montevideo, in Australia, in Panama City, and in grimly embattled England. All the newspapers discussed them without ever suspecting that they had been translated into plain speech. They were featured as freak news—and each new account mentioned that the broadcast reception had ended with a break–down of the receiving apparatus.

Guarded messages passed among the high authorities of the nations that picked up the stuff. A cautious inquiry went even to the Compubs.

The Union of Communist Republics answered characteristically. It asked a question about Mahon units. There were rumors, it said, about a new principle of machine–control lately developed in the United States. It was said that machines equipped with the new units did not wear out, that they exercised seeming intelligence at their tasks, and that they promised to end the enormous drain on natural resources caused by the wearing–out and using–up of standard–type machinery.

The Compub Information Office offered to trade data on the broadcasts for data about the new Mahon–modified machines. It hinted at extremely important revelations it could make.

The rest of the world deduced astutely that the Compubs were scared, too. And they were correct.

* * * * *

Then, quite suddenly, a break came. All previous broadcast receptions had ended with the break–down of the receiving instrument. Now a communicator named Betsy, modified in the Mahon manner and at work in the research installation working with Mahon–modified devices, began to pick up the broadcasts consistently, keeping each one on its screen until it ended.

Day after day, at highly irregular intervals, Betsy's screen lighted up and showed the weird patterns, and her loudspeakers emitted the peepings and chirps and deep–bass hootings of the broadcasts. And the high brass went into a dither to end all dithers as tapes of the received material reached the Pentagon and were translated into intelligible speech and pictures.

* * * * *

This was when Metech Sergeant Bellews, in charge of

the Rehab Shop at Research Installation 83, came into the affair. Specifically, he entered the picture when a young second lieutenant came to the shop to fetch him to Communications Center in that post.

The lieutenant was young and tall and very military. Sergeant Bellews was not. So he snorted, upon receipt of the message. He was at work on a vacuum cleaner at the moment—a Mahon–modified machine with a flickering yellow standby light that wavered between brightness and dimness with much more than appropriate frequency. The Rehabilitation Shop was where Mahon–modified machines were brought back to usefulness when somebody messed them up. Two or three machines—an electric ironer, for one—operated slowly and hesitantly. That was occupational therapy. A washing–machine churned briskly, which was convalescence. Others, ranging from fire–control computers to teletypes and automatic lathes, simply waited with their standby lights flickering meditatively according to the manner and custom of Mahon–modified machines. They were ready for duty again.

The young lieutenant was politely urgent.

"But I been there!" protested Sergeant Bellews. "I checked! It's a communicator I named Betsy. She's all right! She's been mishandled by the kinda halfwits Communications has around, but she's a good, well–

balanced, experienced machine. If she's turning out broadcasts, it's because they're comin' in! She's all right!"

"I know," said the young lieutenant soothingly. His uniform and his manners were beautiful to behold. "But the Colonel wants you there for a conference."

"I got a communicator in the shop here," said Sergeant Bellews suspiciously. "Why don't he call me?"

"Because he wants to try some new adjustments on—ah—Betsy, Sergeant. You have a way with Mahon machines. They'll do things for you they won't do for anybody else."

Sergeant Bellews snorted again. He knew he was being buttered up, but he'd asked for it. He even insisted on it, for the glory of the Metallurgical Technicians' Corps. The big brass tended to regard Metechs as in some fashion successors to the long-vanished veterinary surgeons of the Farriers' Corps, when horses were a part of the armed forces. Mahon-modified machines were new—very new—but the top brass naturally remembered everything faintly analogous and applied it all wrong. So Sergeant Bellews conducted a one-man campaign to establish the dignity of his profession.

But nobody without special Metech training ought to

tinker with a Mahon–modified machine.

"If he's gonna fool with Betsy," said the Sergeant bitterly, "I guess I gotta go over an' boss the job."

He pressed a button on his work–table. The vacuum cleaner's standby light calmed down. The button provided soothing sub–threshold stimuli to the Mahon unit, not quite giving it the illusion of operating perfectly—if a Mahon unit could be said to be capable of illusion—but maintaining it in the rest condition which was the foundation of Mahon–unit operation, since a Mahon machine must never be turned off.

The lieutenant started out of the door. Sergeant Bellews followed at leisure. He painstakingly avoided ever walking the regulation two paces behind a commissioned officer. Either he walked side by side, chatting, or he walked alone. Wise officers let him get away with it.

* * * * *

Reaching the open air a good twenty yards behind the lieutenant, he cocked an approving eye at a police–up unit at work on the lawn outside. Only a couple of weeks before, that unit had been in a bad way. It stopped and shivered when it encountered an unfamiliar object.

But now it rolled across the grass from one path-edge to another. When it reached the second path it stopped, briskly moved itself its own width sidewise, and rolled back. On the way it competently manicured the lawn. It picked up leaves, retrieved a stray cigarette-butt, and snapped up a scrap of paper blown from somewhere. Its tactile units touched a new-planted shrub. It delicately circled the shrub and went on upon its proper course.

* * * * *

Once, where the grass grew taller than elsewhere, it stopped and whirred, trimming the growth back to regulation height. Then it went on about its business as before.

Sergeant Bellews felt a warm sensation. That was a good machine that had been in a bad way and he'd brought it back to normal, happy operation. The sergeant was pleased.

The lieutenant turned into the Communications building. Sergeant Bellews followed at leisure. A jeep went past him—one of the special jeeps being developed at this particular installation—and its driver was talking to someone in the back seat, but the jeep matter-of-factly turned out to avoid Sergeant Bellews. He glowed. He'd activated it. Another good machine, gathering sound experience day by day.

He went into the room where Betsy stood—the communicator which, alone among receiving devices in the whole world, picked up the enigmatic broadcasts consistently. Betsy was a standard Mark IV communicator, now carefully isolated from any aerial. She was surrounded by recording devices for vision and sound, and by the most sensitive and complicated instruments yet devised for the detection of short–wave radiation. Nothing had yet been detected reaching Betsy, but something must. No machine could originate what Betsy had been exhibiting on her screen and emitting from her speakers.

Sergeant Bellews tensed instantly. Betsy's standby light quivered hysterically from bright to dim and back again. The rate of quivering was fast. It was very nearly a sine–wave modulation of the light—and when a Mahon–modified machine goes into sine–wave flicker, it is the same as Cheyne–Stokes breathing in a human.

He plunged forward. He jerked open Betsy's adjustment–cover and fairly yelped his dismay. He reached in and swiftly completed corrective changes of amplification and scanning voltages. He balanced a capacity bridge. He soothed a saw–tooth resonator. He seemed to know by sheer intuition what was needed to be done.

After a moment or two the standby lamp wavered slowly from near–extinction to half–brightness, and then to full brightness and back again. It was completely unrhythmic and very close to normal.

"Who done this?" demanded the sergeant furiously. "He had Betsy close to fatigue collapse! He'd ought to be court–martialed!"

He was too angry to notice the three civilians in the room with the colonel and the lieutenant who'd summoned him. The young officer looked uncomfortable, but the colonel said authoritatively:

"Never mind that, Sergeant. Your Betsy was receiving something. It wasn't clear. You had not reported, as ordered, so an attempt was made to clarify the signals."

"Okay, Colonel!" said Sergeant Bellews bitterly. "You got the right to spoil machines! But if you want them to work right you got to treat 'em right!"

"Just so," said the colonel. "Meanwhile—this is Doctor Howell, Doctor Graves, and Doctor Lecky. Sergeant Bellews, gentlemen. Sergeant, these are not MDs. They've been sent by the Pentagon to work on Betsy."

* * * * *

"Betsy don't need workin' on!" said Sergeant Bellews belligerently. "She's a good, reliable, experienced machine! If she's handled right, she'll do better work than any machine I know!"

"Granted," said the colonel. "She's doing work now that no other machine seems able to do—drawing scrambled broadcasts from somewhere that can only be guessed at. They've been unscrambled and these gentlemen have come to get the data on Betsy. I'm sure you'll cooperate."

"What kinda data do they want?" demanded Bellews. "I can answer most questions about Betsy!"

"Which," the colonel told him, "is why I sent for you. These gentlemen have the top scientific brains in the country, Sergeant. Answer their questions about Betsy and I think some very high brass will be grateful.

"By the way, it is ordered that from now on no one is to refer to Betsy or any work on these broadcasts, over any type of electronic communication. No telephone, no communicator, no teletype, no radio, no form of communication except viva voce. And that means you talking to somebody else, Sergeant, with no microphone around. Understand? And from now on you will not talk about anything at all except to these gentlemen and to me."

Sergeant Bellews said incredulously:

"Suppose I got to talk to somebody in the Rehab Shop. Do I signal with my ears and fingers?"

"You don't talk," said the colonel flatly. "Not at all."

Sergeant Bellews shook his head sadly. He regarded the colonel with such reproach that the colonel stiffened. But Sergeant Bellews had a gift for machinery. He had what amounted to genius for handling Mahon–modified devices. So long as no more competent men turned up, he was apt to get away with more than average.

The colonel frowned and went out of the room. The tall young lieutenant followed him faithfully. The sergeant regarded the three scientists with the suspicious air he displayed to everyone not connected with Mahon units in some fashion.

"Well?" he said with marked reserve. "What can I tell you first?"

Lecky was the smallest of the three scientists. He said ingratiatingly, with the faintest possible accent in his speech:

"The nicest thing you could do for us, Sergeant, would

be to show us that this—Betsy, is it?—with other machines before her, has developed a contagious machine insanity. It would frighten me to learn that machines can go mad, but I would prefer it to other explanations for the messages she gives."

"Betsy can't go crazy," said Bellews with finality. "She's Mahon–controlled, but she hasn't got what it takes to go crazy. A Mahon unit fixes a machine so it can loaf and be a permanent dynamic system that can keep acquired habits of operatin'. It can take trainin'. It can get to be experienced. It can learn the tricks of its trade, so to speak. But it can't go crazy!"

"Too bad!" said Lecky. He added persuasively: "But a machine can lie, Sergeant? Would that be possible?"

Sergeant Bellews snorted in denial.

* * * * *

"The broadcasts," said Lecky mildly, "claim a remarkable reason for certainty about an extremely grave danger which is almost upon the world. If it's the truth, Sergeant, it is appalling. If it is a lie, it may be more appalling. The Joint Chiefs of Staff take it very seriously, in any case. They—"

"I got cold shivers," said Sergeant Bellews with irony.

"I'm all wrought up. Huh! The big brass gets the yellin' yollups every so often anyhow. Listen to them, and nothin' happens except it's top priority top secret extra crash emergency! What do you want to know about Betsy?"

There was a sudden squealing sound from the communicator on which all the extra recording devices were focussed. Betsy's screen lighted up. Peculiarly curved patterns appeared on it. They shifted and changed. Noises came from her speaker. They were completely unearthly. Now they were shrill past belief, and then they were chopped into very small bits of sound, and again they were deepest bass, when each separate note seemed to last for seconds.

"You might," said Lecky calmly, "tell us from where your Betsy gets the signal she reports in this fashion."

There were whirrings as recorders trained upon Betsy captured every flickering of her screen and every peeping noise or deep–toned rumble. The screen–pattern changed with the sound, but it was not linked to it. It was a completely abnormal reception. It was uncanny. It was somehow horrible because so completely remote from any sort of human communication in the year 1972.

The three scientists watched with worried eyes. A

communicator, even with a Mahon unit in it, could not originate a pattern like this! And this was not conceivably a distortion of anything transmitted in any normal manner in the United States of America, or the Union of Compubs, or any of the precariously surviving small nations not associated with either colossus.

"This is a repeat broadcast!" said one of the three men suddenly. It was Howell, the heavy-set man. "I remember it. I saw it projected—like this, and then unscrambled. I think it's the one where the social system's described—so we can have practice at trying to understand. Remember?"

* * * * *

Lecky said, as if the matter had been thrashed out often before:

"I do not believe what it says, Howell! You know that I do not believe it! I will not accept the theory that this broadcast comes from the future!"

The broadcast stopped. It stopped dead. Betsy's screen went blank. Her wildly fluctuating standby light slowed gradually to a nearly normal rate of flicker.

"That's not a theory," said Howell dourly. "It's a statement in the broadcast. We saw the first transmission of this from the tape at the Pentagon. Then we saw it with the high–pitched parts slowed down and the deep–bass stuff speeded up. Then it was a human voice giving data on the scanning pattern and then rather drearily repeating that history said that intertemporal communication began with broadcasts sent back from 2180 to 1972. It said the establishment of two–way communication was very difficult and read from a script about social history, to give us practice in unscrambling it. It's not a theory to say the stuff originates in the future. It's a statement."

"Then it is a lie," said Lecky, very earnestly. "Truly, Howell, it is a lie!"

"Then where does the broadcast come from?" demanded Howell. "Some say it's a Compub trick. But if they were true they'd hide it for use to produce chaos in a sneak attack. The only other theory—"

* * * * *

Graves, the man with the short moustache, said jerkily:

"No, Howell! It is not an extra–terrestrial creature pretending to be a man of our own human future. One could not sleep well with such an idea in his head. If some non–human monster could do this—"

"I do not sleep at all," said Lecky simply. "Because it says that two-way communication is to come. I can listen to these broadcasts tranquilly, but I cannot bear the thought of answering them. That seems to me madness!"

Sergeant Bellews said approvingly:

"You got something there! Yes, sir! Did you notice how Betsy's standby light was wabbling while she was bringin' in that broadcast? If she could sweat, she'd've been sweating!"

Lecky turned his head to stare at the sergeant.

"Machines," said Bellews profoundly, "act according to the golden rule. They do unto you as they would have you do unto them. You treat a machine right and it treats you right. You treat it wrong and it busts itself—still tryin' to treat you right. See?"

Lecky blinked.

"I do not quite see how it applies," he said mildly.

"Betsy's an old, experienced machine," said the sergeant. "A signal that makes her sweat like that has got something wrong about it. Any ordinary machine 'ud break down handlin' it."

Graves said jerkily:

"The other machines that received these broadcasts did break down, Sergeant. All of them."

"Sure!" said the sergeant with dignity. "O' course, who's broadcastin' may have been tinkerin' with their signal since they seen it wasn't gettin' through. Betsy can take it now, when younger machines with less experience can't. Maybe a micro–microwatt of signal. Then it makes her sweat. If she was broadcastin', with a hell of a lot more'n a micro–microwatt—it'd be bad! I bet you that every machine we make to broadcast breaks down! I bet—"

Howell said curtly:

"Reasonable enough! A signal to pass through time as well as space would be different from a standard wave–type! Of course that must be the answer."

Sergeant Bellews said truculently:

"I got a hunch that whoever's broadcastin' is busting transmitters right an' left. I never knew anything about this before, except that Betsy was pickin' up stuff that came from nowhere. But I bet if you look over the record–tapes you will find they got breaks where one transmitter switched off or broke down and another took over!"

Lecky's eyes were shining. He regarded Sergeant Bellews with a sort of tender respect.

"Sergeant Bellews," he said softly, "I like you very much. You have told us undoubtedly true things."

"Think nothin' of it," said the sergeant, gratified. "I run the Rehab Shop here, and I could show you things—"

"We wish you to," said Lecky. "The reaction of machines to these broadcasts is the one viewpoint we would never have imagined. But it is plainly important. Will you help us, Sergeant? I do not like to be frightened—and I am!"

"Sure, I'll help," said Sergeant Bellews largely. "First thing is to whip some stuff together so we can find out what's what. You take a few Mahon units, and install 'em and train 'em right, and they will do almost anything you've a mind for. But you got to treat 'em right. Machines work by the golden rule. Always! Come along!"

* * * * *

Sergeant Bellews went to the Rehab Shop, followed only by Lecky. All about, the sun shone down upon buildings with a remarkably temporary look about them, and on lawns with a remarkably lush look about them, and signboards with very black lettering on gray paint backgrounds. There was a very small airfield inside the barbed–wire fence about the post, and elaborate machine–shops, and rows and rows of barracks and a canteen and a USO theatre, and a post post–office. Everything seemed quite matter–of–fact.

Except for the machines.

They were the real reason for the existence of the post. The barracks and married–row dwellings had washing–machines which looked very much like other washing–machines, except that they had standby lights which flickered meditatively when they weren't being used.

* * * * *

The television receivers looked like other TV sets, except for minute and wavering standby lights which were never quite as bright or dim one moment as the next. The jeeps—used strictly within the barbed–wire fence around the post—had similar yellow glowings on their instrument–boards, and they were very remarkable jeeps. They never ran off the graveled roads onto the grass, and they never collided with each other, and it was said that the nine–year–old son of a lieutenant–colonel had tried to drive one and it would not stir. Its motor cut off when he forced it into gear. When he tried to re–start it, the starter did not turn. But when an adult stepped into it, it operated perfectly—only it braked and stopped itself when a small child toddled into its path.

There were some people who said that this story was not true, but other people insisted that it was. Anyhow the washing–machines were perfect. They never tangled clothes put into them. It was reported that Mrs. So–and–so's washing–machine had found a load of clothes tangled, and reversed itself and worked backward until they were straightened out.

Television sets turned to the proper channels—different ones at different times of day—with incredible facility. The smallest child could wrench at a tuning–knob and the desired station came on. All the operating devices of Research Installation 83 worked as if they liked to—which might have been alarming except that they never did anything of themselves. They initiated nothing. But each one acted like an old, favorite possession. They fitted their masters. They seemed to tune themselves to the habits of their owners. They were infinitely easy to work right, and practically impossible to work wrong.

Such machines, of course, had not been designed to cope with enigmatic broadcasts or for military purposes. But the jet–planes on the small airfield were very remarkable indeed, and the other and lesser devices had been made for better understanding of the Mahon units which made machines into practically a new order of creation.

* * * * *

Sergeant Bellews ushered Lecky into the Rehab Shop. There was the pleasant, disorderly array of devices with their wavering standby lights. They gave an effect of being alive, but somehow it was not disturbing. They seemed not so much intent as meditative, and not so much watchful as interested. When the sergeant and his guest moved past them, the unrhythmic waverings of the small yellow lights seemed to change hopefully, as if the machines anticipated being put to use. Which, of course, was absurd. Mahon machines do not anticipate anything. They probably do not remember anything, though patterns of operation are certainly retained in very great variety. The fact is that a Mahon unit is simply a device to let a machine stand idle without losing the nature of an operating machine.

The basic principle goes back to antiquity. Ships, in ancient days, had manners and customs individual to each vessel. Some were sweet craft, easily handled and staunch and responsive. Others were stubborn and begrudging of all helpfulness. Sometimes they were even man–killers. These facts had no rational explanation, but they were facts. In similarly olden times, particular weapons acquired personalities to the point of having personal names—Excalibur, for example.

Every fighting man knew of weapons which seemed to possess personal skill and ferocity. Later, workmen found that certain tools had a knack of fitting smoothly in the hand—seeming even to divine the grain of the wood they worked on. The individual characteristics of violins were notorious, so that a violin which sang joyously under the bow was literally priceless.

And all these things, as a matter of observation and not of superstition, kept their qualities only when in constant use. Let a ship be hauled out of water and remain there for a time, and she would be clumsy on return to her native element. Let a sword or tool stay unused, and it seemed to dull. In particular, the finest of violins lost its splendor of tone if left unplayed, and any violin left in a repair–shop for a month had to be played upon constantly for many days before its living tone came back.

* * * * *

The sword and the tool perhaps, but the ship and the violin certainly, acted as if they acquired habits of operation by being used, and lost them by disuse. When more complex machines were invented, such facts were less noticeable. True, no two automobiles ever handled exactly the same, and that was recognized. But the fact that no complex machine worked well until it had run for a time was never commented on, except in the observation that it needed to be warmed up. Anybody would have admitted that a machine in the act of operating was a dynamic system in a solid group of objects, but nobody reflected that a stopped machine was a dead thing. Nobody thought to liken the warming–up period for an aeroplane engine to the days of playing before a disuse–dulled violin regained its tone.

Yet it was obvious enough. A ship and a sword and a tool and a violin were objects in which dynamic systems existed when they were used, and in which they ceased to exist when use stopped. And nobody noticed that a living creature is an object which contains a dynamic system when it is living, and loses it by death.

For nearly two centuries quite complex machines were started, and warmed up, and used, and then allowed to grow cold again. In time the more complex machines were stopped only reluctantly. Computers, for example, came to be merely turned down below operating voltage when not in use, because warming them up was so difficult and exacting a task. Which was an unrecognized use of the Mahon principle. It was a way to keep a machine activated while not actually operating. It was a state of rest, of loafing, of idleness, which was not the death of a running mechanism.

The Mahon unit was a logical development. It was an absurdly simple device, and not in the least like a brain. But to the surprise of everybody, including its inventor, a Mahon–modified machine did more than stay warmed up. It retained operative habits as no complex device had ever done before. In time it was recognized that Mahon–modified machines acquired experience and kept it so long as the standby light glowed and flickered in its socket. If the lamp went out the machine died, and when reënergized was a different individual entirely, without experience.

Sergeant Bellews made such large–minded statements as were needed to brief Lecky on the work done in this installation with Mahon–controlled machines.

"They don't think," he explained negligently, "any more than dogs think. They just react—like dogs do. They get patterns of reaction. They get trained. Experienced. They get good! Over at the airfield they're walking around beaming happy over the way the jets are flyin' themselves."

Lecky gazed around the Rehab Shop. There were shelves of machines, duly boxed and equipped with

Mahon units, but not yet activated. Activation meant turning them on and giving them a sort of basic training in the tasks they were designed to do. But also there were machines which had broken down—invariably through misuse, said Sergeant Bellews acidly—and had been sent to the Rehab Shop to be re-trained in their proper duties.

"Guys see 'em acting sensible and obediently," said Bellews with bitterness, "and expect 'em to think. Over at the jet-field they finally come to understand." His tone moderated. "Now they got jets that put down their own landing-gear, and holler when fuel's running low, and do acrobatics happy if you only jiggle the stick. They mighty near fly themselves! I tell you, if well-trained Mahon jets ever do tangle with old-style machines, it's goin' to be a caution to cats! It'll be like a pack of happy terriers pilin' into hamsters. It'll be murder!"

* * * * *

He surveyed his stock. From a back corner he brought out a small machine with an especially meditative tempo in its standby-lamp flicker. The tempo accelerated a little when he put it on a work-bench.

"They got batteries to stay activated with," he observed, "and only need real juice when they're workin'. This here's a play–back recorder they had over in Recreation. Some guys trained it to switch frequencies —speed–up and slow–down stuff. They laughed themselves sick! There used to be a tough guy over there,—a staff sergeant, he was—that gave lectures on military morals in a deep bass voice. He was proud of that bull voice of his. He used it frequently. So they taped him, and Al here—" the name plainly referred to the machine—"used to play it back switched up so he sounded like a squeaky girl. That poor guy, he liked to busted a blood–vessel when he heard himself speakin' soprano. He raised hell and they sent Al here to be rehabilitated. But I switched another machine for him and sent it back, instead. Of course, Al don't know what he's doing, but—"

* * * * *

He brought over another device, slightly larger and with a screen.

"Somebody had a bright notion with this one, too," he said. "They figured they'd scramble pictures for secret transmission, like they scramble voice. But they found they hadda have team–trained sets to work, an' they weren't interchangeable. They sent Gus here to be deactivated an' trained again. I kinda hate to do that. Sometimes you got to deactivate a machine, but it's like shooting a dog somebody's taught to steal eggs, who don't know it's wrong."

He bolted the two instruments together. He made connections with flexible cables and tucked the cable out of sight. He plugged in for power and began to make adjustments.

The small scientist asked curiously:

"What are you preparing, Sergeant?"

"These two'll unscramble that broadcast," said Sergeant Bellews, with tranquil confidence. "Al's learned how to make a tape and switch frequencies expert. Gus, here, he's a unscrambler that can make any kinda scanning pattern. Together they'll have a party doing what they're special trained for. We'll hook 'em to Betsy's training–terminals."

He regarded the two machines warmly. Connected, now, their standby lights flickered at a new tempo. They synchronized, and broke synchrony, and went back into elaborate, not–quite–resolvable patterns which were somehow increasingly integrated as seconds went by.

"Those lights look kinda nice, don't they?" asked the sergeant admiringly. "Makes you think of a coupla dogs gettin' acquainted when they're goin' out on a job of hunting or something."

But Lecky said abruptly, in amazement:

"But, Sergeant! In the Pentagon it takes days to unscramble a received broadcast such as Betsy receives! It requires experts—"

"Huh!" said Sergeant Bellews. He picked up the two machines. "Don't get me started about the kinda guys that wangle headquarters–company jobs! They got a special talent for fallin' soft. But they haven't necessarily got anything else!"

* * * * *

Lecky followed Sergeant Bellews as the sergeant picked up his new combination of devices and headed out of the Rehab Shop. Outside, in the sunshine, there were roarings to be heard. Lecky looked up. A formation of jets swam into view against the sky. A tiny speck, trailing a monstrous plume of smoke, shot upward from the jet–field. The formation tightened.

The ascending jet jiggled as if in pure exuberance as it swooped upward—but the jiggle was beautifully designed to throw standard automatic gunsights off.

A plane peeled off from the formation and dived at the ascending ship. There was a curious alteration in the thunder of motors. The rate–of–rise of the climbing jet dwindled almost to zero. Sparks shot out before it. They made a cone the diving ship could not avoid. It sped through them and then went as if disappointedly to a lower level. It stood by to watch the rest of the dog-fight.

"Nice!" said Sergeant Bellews appreciatively. "That's a Mahon jet all by itself, training against regular ships. They have to let it shoot star-bullets in training, or it'd get hot and bothered in a real fight when its guns went off."

The lower jet streaked skyward once more. Sparks sped from the formation. They flared through emptiness where the Mahon jet had been but now was not. It scuttled abruptly to one side as concerted streams of sparks converged. They missed. It darted into zestful, exuberant maneuverings, remarkably like a dog dashing madly here and there in pure high spirits. The formation of planes attacked it resolutely.

Suddenly the lone jet plunged into the midst of the formation, there were coruscations of little shooting stars, and one-two-three planes disgustedly descended to lower levels as out of action. Then the single ship shot upward, seemed eagerly to shake itself, plunged back—and the last ships tried wildly to escape, but each in turn was technically shot down.

The Mahon jet headed back for its own tiny airfield. Somehow, it looked as if, had it been a dog, it would be wagging its tail and panting happily.

"That one ship," said Lecky blankly, "it defeated the rest?"

"It's got a lot of experience," said the sergeant. "You can't beat experience."

He led the way into Communications Center. In the room where Betsy stood, Howell and Graves had been drawing diagrams at each other to the point of obstinacy.

"But don't you see?" insisted Howell angrily. "There can be no source other than a future time! You can't send short waves through three–dimensional space to a given spot and not have them interceptible between. Anyhow, the Compubs wouldn't work it this way! They wouldn't put us on guard! And an extra–terrestrial wouldn't pretend to be a human if he honestly wanted to warn us of danger! He'd tell us the truth! Physically and logically it's impossible for it to be anything but what it claims to be!"

Graves said doggedly:

"But a broadcast originating in the future is

impossible!"

"Nothing," snapped Howell, "that a man can imagine is impossible!"

"Then imagine for me," said Graves, "that in 2180 they read in the history books about a terrible danger to the human race back in 1972, which was averted by a warning they sent us. Then, from their history–books, which we wrote for them, they learn how to make a transmitter to broadcast back to us. Then they tell us how to make a transmitter to broadcast ahead to them. They don't invent the transmitter. We tell them how to make it—via a history book. We don't invent it. They tell us—from the history book. Now imagine for me how that transmitter got invented!"

"You're quibbling," snapped Howell. "You're refusing to face a fact because you can't explain it. I say face the fact and then ask for an explanation!"

"Why not ask them," said Graves, "how to make a round square or a five–sided triangle?"

* * * * *

Sergeant Bellews pushed to a spot near Betsy. He put down his now–linked Mahon machines and began to move away some of the recording apparatus focused on

Betsy.

"Hold on there!" said Howell in alarm. "Those are recorders!"

"We'll let 'em record direct," said the sergeant.

* * * * *

Lecky spoke feverishly in support of Bellews. But what he said was, in effect, a still–marveling description of the possibilities of Mahon–modified machines. They were, he said with ardent enthusiasm, the next step in the historic process by which successively greater portions of the cosmos enter into a symbiotic relationship with man. Domestic animals entered into such a partnership aeons ago. Certain plants—wheat and the like—even became unable to exist without human attention. And machines were wrought by man and for a long time served him reluctantly. Pre–Mahon machines were tamed, not domestic. They wore themselves out and destroyed themselves by accidents. But now there were machines which could enter into a truly symbiotic relationship with humanity.

"What," demanded Howell, "what in hell are you talking about?"

Lecky checked himself. He smiled abashedly:

"I think," he said humbly, "that I speak of the high destiny of mankind. But the part that applies at the moment is that Sergeant Bellews must not be interfered with."

He turned and ardently assisted Sergeant Bellews in making room for the just-brought devices. Sergeant Bellews led flexible cables from them to Betsy. He inserted their leads in her training-terminals. He made adjustments within.

It became notable that Betsy's standby light took up new tempos in its wavering. There were elaborate interweavings of rate and degree of brightening among the lights of all three instruments. There was no possible way to explain the fact, but a feeling of pleasure, of zestful stirring, was somehow expressed by the three machines which had been linked together into a cooperating group.

Sergeant Bellews eased himself into a chair.

"Now everything's set," he observed contentedly. "Remember, I ain't seen any of these broadcasts unscrambled. I don't know what it's all about. But we got three Mahon machines set up now to work on the next crazy broadcast that comes in. There's Betsy and these two others. And all machines work accordin' to the Golden Rule, but Mahon machines—they are honey–babes! They'll bust themselves tryin' to do what you ask 'em. And I asked these babies for plenty—only not enough to hurt 'em. Let's see what they turn out."

He pulled a pipe and tobacco from his pocket. He filled the pipe. He squeezed the side of the bowl and puffed as the tobacco glowed. He relaxed, underneath the wall–sign which sternly forbade smoking by all military personnel within these premises.

It was nearly three hours—but it could have been hundreds—before Betsy's screen lighted abruptly.

* * * * *

The broadcast came in; a new transmission. The picture–pattern on Betsy's screen was obviously not the same as other broadcasts from nowhere. The chirps and peepings and the rumbling deep sounds were not repetitions of earlier noise–sequences. It should have taken many days of finicky work by technicians at the Pentagon before the originally broadcast picture could be seen and the sound interpreted. But a play–back recorder named Al, and a picture–unscrambler named Gus were in closed–circuit relationship with Betsy. She received the broadcast and they unscrambled the sound and vision parts of it immediately.

The translated broadcast, as Gus and Al presented it,

was calculated to put the high brass of the defense forces into a frenzied tizzy. The anguished consternation of previous occasions would seem like very calm contemplation by comparison. The high brass of the armed forces should grow dizzy. Top–echelon civilian officials should tend to talk incoherently to themselves, and scientific consultants—biologists in particular—ought to feel their heads spinning like tops.

The point was that the broadcast had to be taken seriously because it came from nowhere. There was no faintest indication of any signal outside of Betsy's sedately gray–painted case. But Betsy was not making it up. She couldn't. There was a technology involved which required the most earnest consideration of the message carried by it.

And this broadcast explained the danger from which the alleged future wished to rescue its alleged past. A brisk, completely deracialized broadcaster appeared on Gus's screen.

In clipped, oddly stressed, but completely intelligible phrases, he explained that he recognized the paradox his communication represented. Even before 1972, he observed, there had been argument about what would happen if a man could travel in time and happened to go back to an earlier age and kill his grandfather. This communication was an inversion of that paradox. The world of 2180 wished to communicate back in time and save the lives of its great–great–great–grandparents so that it—the world of 2180—would be born.

Without this warning and the information to be given, at least half the human race of 1972 was doomed.

In late 1971 there had been a mutation of a minor strain of staphylococcus somewhere in the Andes. The new mutation thrived and flourished. With the swift transportation of the period, it had spread practically all over the world unnoticed, because it produced no symptoms of disease.

Half the members of the human race were carriers of the harmless mutated staphylococcus now, but it was about to mutate again in accordance with Gordon's Law (the reference had no meaning in 1972) and the new mutation would be lethal. In effect, one human being in two carried in his body a semi–virus organization which he continually spread, and which very shortly would become deadly. Half the human race was bound to die unless it was instructed as to how to cope with it. Unless—

*****

Unless the world of 2180 told its ancestors what to do about it. That was the proposal. Two–way communication was necessary for the purpose, because there would be questions to be answered, obscure points to be clarified, numerical values to be checked to the highest possible degree of accuracy.

Therefore, here were diagrams of the transmitter needed to communicate with future time. Here were enlarged diagrams of individual parts. The enigmatic parts of the drawing produced a wave–type unknown in 1972. But a special type of wave was needed to travel beyond the three dimensions of ordinary space, into the fourth dimension which was time. This wave–type produced unpredictable surges of power in the transmitter, wherefore at least six transmitters should be built and linked together so that if one ceased operation another would instantly take up the task.

* * * * *

The broadcast ended abruptly. Betsy's screen went blank. The colonel was notified. A courier took tapes to Washington by high–speed jet. Life in Research Establishment 83 went on sedately. The barracks and the married quarters and the residences of the officers were equipped with Mahon–modified machines which laundered diapers perfectly, and with dial telephones which always rang right numbers, and there were police–up machines which took perfect care of lawns, and television receivers tuned themselves to the customary channels for different hours with astonishing ease. Even jet-planes equipped with Mahon units almost landed themselves, and almost flew themselves about the sky in simulated combat with something very close to zest.

But the atmosphere in the room in Communications was tense.

"I think," said Howell, with his lips compressed, "that this answers all your objections, Graves. Motive—"

"No," said Lecky painfully. "It does not answer mine. My objection is that I do not believe it."

"Huh!" said Sergeant Bellews scornfully. "O' course, you don't believe it! It's phoney clear through!"

Lecky looked at him hopefully.

"You noticed something that we missed, Sergeant?"

"Hell, yes!" said Sergeant Bellews. "That transmitter diagram don't have a Mahon unit in it!"

"Is that remarkable?" demanded Howell.

"Remarkable dumb," said the sergeant. "They'd ought to know—"

The tall young lieutenant who earlier had fetched Sergeant Bellews to Communications now appeared again. He gracefully entered the room where Betsy waited for more broadcast matter. Her standby light flickered with something close to animation, and the similar yellow bulbs on Al and Gus responded in kind. The tall young lieutenant said politely:

"I am sorry, but pending orders from the Pentagon the colonel has ordered this room vacated. Only automatic recorders will be allowed here, and all records they produce will be sent to Washington without examination. It seems that no one on this post has the necessary clearance for this type of material."

Lecky blinked. Graves sputtered:

"But—dammit, do you mean we can work out a way to receive a broadcast and not be qualified to see it?"

"There's a common–sense view," said Sergeant Bellews oracularly, "and a crazy view, and there's what the Pentagon says, which ain't either." He stood up. "I see where I go back to my shop and finish rehabilitatin' the colonel's vacuum cleaner. You gentlemen care to join me?"

Howell said indignantly:

"This is ridiculous! This is absurd!"

"Uh–uh," said Sergeant Bellews benignly. "This is the armed forces. There'll be an order makin' some sort of sense come along later. Meanwhile, I can brief you guys on Mahon machines so you'll be ready to start up again with better information when a clearance order does come through. And I got some beer in my quarters behind the Rehab Shop. Come along with me!"

He led the way out of the room. The young lieutenant paused to close the door firmly behind him and to lock it. A bored private, with side–arms, took post before it. The lieutenant was a very conscientious young man.

But he did not interfere with the parade to Sergeant Bellews' quarters. The young lieutenant was very military, and the ways of civilians were not his concern. If eminent scientists chose to go to Sergeant Bellews' quarters instead of the Officers Club, to which their assimilated rank entitled them, it was strictly their affair.

* * * * *

They reached the Rehab Shop, and Sergeant Bellews went firmly to a standby–light–equipped refrigerator in his quarters. He brought out beer and deftly popped off the tops. The icebox door closed quietly.

"Here's to crime," said Sergeant Bellews amiably.

He drank. Howell sipped gloomily. Graves drank thoughtfully. Lecky looked anticipative.

"Sergeant," he said, "did I see a gleam in your eye just now?"

Sergeant Bellews reflected, gently shaking his opened beer–can with a rotary motion, for no reason whatever.

"Uh–uh," he rumbled. "I wouldn't say a gleam. But you mighta seen a glint. I got some ideas from what I seen during that broadcast. I wanna get to work on 'em. Here's the place to do the work. We got facilities here."

Howell said with precise hot anger:

"This is the most idiotic situation I have ever seen even in government service!"

"You ain't been around much," the sergeant told him kindly. "It happens everywhere. All the time. It ain't even a exclusive feature of the armed forces." He put down his beer–can and patted his stomach. "There's guys who sit up nights workin' out standard operational procedures just to make things like this happen, everywhere. The colonel hadda do what he did. He's got orders, too. But he felt bad. So he sent the lieutenant to tell us. He does the colonel's dirty jobs—and he loves his work."

* * * * *

He moved grandly toward the Rehab Shop proper, which opened off the quarters he lived in—very much as a doctor's office is apt to open off his living quarters.

"We follow?" asked Lecky zestfully. "You plan something?"

"Natural!" said Sergeant Bellews largely.

He led the way into the Rehab Shop, which was dark and shadowy, and only very dimly lighted by flickering, wavering lights of many machines waiting as if hopefully to be called on for action. There were the shelves of machines not yet activated. Sergeant Bellews led the way toward his desk. There was a vacuum cleaner on it, on standby. He put it down on the floor.

Lecky watched him with some eagerness. The others came in, Howell dourly and Graves wiping his moustache.

The sergeant considered his domain.

"We'll be happy to help you," said Lecky.

"Thanks," said the sergeant. "I'm under orders to help you, too, y'know. Just supposing you asked me to whip up something to analyze what Betsy receives, so it can be checked on that it is a new wave–type."

"Can you do that?" demanded Graves. "We were supposed to work on that—but so far we've absolutely nothing to go on!"

The sergeant waved his hand negligently.

"You got something now. Betsy's a Mahon–modified device. Every receiver that picked up one of those crazy broadcasts broke down before it was through. She takes 'em in her stride—especial with Al and Gus to help her. Wouldn't it be reasonable to guess that Mahon machines are—uh—especial adapted to handle intertemporal communication?"

"Very reasonable!" said Howell dourly. "Very! The broadcast said that the wave–type produced unpredictable surges of current. Ordinary machines do find it difficult to work with whatever type of radiation that can be."

"Betsy chokes off those surges," observed the sergeant. "With Gus and Al to help, she don't have no trouble. We hadn't ought to need to make any six transmitters if we put Mahon–unit machines together for the job!"

"Quite right," agreed Lecky, mildly. "And it is odd—"

"Yeah," said the sergeant. "It's plenty odd my great–great–great–grandkids haven't got sense enough to do it themselves!"

* * * * *

He went to a shelf and brought down a boxed machine, —straight from the top–secret manufactory of Mahon units. It had never been activated. Its standby light did not glow. Sergeant Bellews ripped off the carton and said reflectively:

"You hate to turn off a machine that's got its own ways of working. But a machine that ain't been activated has not got any personality. So you don't mind starting it up to turn it off later."

He opened the adjustment–cover and turned something on. The standby light glowed. Closely observed, it was not a completely steady glow. There were the faintest possible variations of brightness. But there was no impression of life.

Graves said:

"Why doesn't it flicker like the others?"

"No habits," said the sergeant. "No experience. It's like

a newborn baby. It'll get to have personality after it's worked a while. But not now."

He went across the shop again. He moved out a heavy case, and twisted the release, and eased out a communicator of the same type—Mark IV—as Betsy back in the Communications room. Howell went to help him. Graves tried to assist. Lecky moved other things out of the way. They were highly eminent scientists, and Metech Sergeant Bellews was merely a non–commissioned officer in the armed forces. But he happened to have specialized information they had not. Quite without condescension they accepted his authority in his own field, and therefore his equality. As civilians they had no rank to maintain, and they disagreed with each other—and would disagree with the sergeant—only when they knew why. Which was one of the reasons why they were eminent scientists.

Sergeant Bellews brought out yet another box. He unrolled cables. He selected machines whose flickering lights seemed to bespeak eagerness to be of use. He coupled them to the newly unboxed machines, whose lights were vaguely steady.

"Training cables," he said over his shoulder. "You get one machine working right, and you hook it with another, and the new machine kinda learns from the old one. Kinda! But it ain't as good as real experience. Not at first."

* * * * *

Presently the lights of the newly energized machines began to waver in somewhat the manner of the ready–for–operation ones. But they did not give so clear an impression of personality.

"Look!" said Sergeant Bellews abruptly. "I got to check with you. The more I think, the more worried I get."

"You begin to believe the broadcasts come from the future?" demanded Graves. "And it worries you? But they do not speak of Mahon units—"

"I don't care where they come from," said the sergeant. "I'm worryin' about what they are! The guy in the broadcast—not knowing Mahon units—said we'd have to make half a dozen transmitters so they'd take over one after another as they blew out. You see what that means?"

Lecky said crisply:

"You pointed it out before. There is something in the wave–type which—you would say this, Sergeant!—which machines do not like. Is that the reasoning?"

"Uh–uh!" The sergeant scowled. "Machines work by the

golden rule. They try to do unto you what they want you to do unto them. Likes an' dislikes don't matter. I mean that there's something about that wave–type that machines can't take! It busts them. If it sort of explodes surges of current in 'em—Look! Any running machine is a dynamic system in a object. A jet–plane operating is that. So's a water–spout. So's a communicator. But if you explode surges of heavy current in a dynamic system in a operating machine—things get messed up. The operating habit is busted to hell. I'm saying that if this wave–type makes crazy surges of current start up— why—if the surges are strong enough they'll bust not only a communicator but a jet–plane. Or a water–spout. Anything! See?"

* * * * *

Lecky blinked and suddenly went pale.

"But," said Howell reasonably, "you said that Betsy handled it. Especially well when linked with other Mahon machines."

"Yeah," said the sergeant.

"I think," observed Graves jerkily, "that you are preparing new machines, without developed— personalities, because you think that if they make this special–type wave they'll be broken."

"Yeah," said the sergeant, again. "The signal Betsy was amplifyin' coulda been as little as a micro–micro–watt. At its frequency an' type, she'd choke it down if it was more. But even a micro–micro–watt bothered Betsy until she got Al and Gus to help. She was fair screamin' for somebody to come help her hold it. But the three of them done all right."

Howell conceded the point.

"That seems sound reasoning."

"But you don't broadcast with a micro–micro–watt. You use a hell of a lot more power than that! The transmitter the guy in the screen said to make was a twenty-kilowatt job. Not too much for a broadcast of sine waves, but a hell of a lot to be turned loose, in waves that have Betsy hollerin' at the power she was handlin'!"

"It might break even the Mahon machines in this installation?" demanded Howell.

"You're gettin' warm," said the sergeant.

Graves said:

"You mean it might break all operating communicators in a very large area?"

"You're gettin' hot," said the sergeant grimly.

Lecky wetted his lips.

"I think," he said very carefully, "that you suspect it is a wave–type which will break any dynamic system, in any sort of object a dynamic system can exist in."

"Yeah," said the sergeant. He waited, looking at Lecky.

"And," said Lecky, "not only operating machines are dynamic systems. Living plants and animals are, too. So are men."

"That's what I'm drivin' at," said Sergeant Bellews.

"So you believe," said Lecky, very pale indeed, "that we have been given the circuit–diagram of a transmitter which will broadcast a wave–type which destroys dynamic systems—life as well as the operation of machines. Persons—in the future or an alien creature in a space–ship, or perhaps even the Compubs—are furnishing us with designs for transmitters of death, to be linked together so that if one fails the others will carry on. And they lure us to destroy ourselves by lying about who they are and what they propose."

"They're lyin'," said the sergeant. "They say they're in the future and they don't know a thing about Mahon units. Else they'd use 'em."

Lecky wetted his lips again.

"And—if they are not in the future, they are trying to get us to destroy ourselves because that would be safer and surer than trying to destroy us by—say—transmitters of death dropped upon us by parachute. Yet if we do not destroy ourselves, they will surely do that."

"If we don't bump ourselves off, it'll be because we got wise," acknowledged the sergeant. "If we get wise, we could bump them off by parachute–transmitter. So they'll beat us to it. They'll have to!"

"Yes," said Lecky. "They'll have to. It has always been said that a death–ray was impossible. This would be a death–broadcast. If we do not broadcast, they will—whoever they are. It is—" He smiled mirthlessly at the magnitude of his understatement. "It is urgent that we do something. What shall we do, Sergeant?"

*****

A squadron of light tanks arrived at Research Installation 83 that afternoon, with a shipment of courier motorcycles. They had been equipped with Mahon units and went to the post to be trained.

The Pentagon was debating the development of a Mahon–modified guided missile, and a drone plane was under construction. But non–military items also arrived for activation and test. Automatic telephone switching systems, it appeared, could be made much simpler if they could be trained to do their work instead of built so they couldn't help it.

Passenger–cars other than jeeps showed promise. It had long been known that most accidents occurred with new cars, and that ancient jalopies were relatively safe even in the hands of juvenile delinquents. It was credible that part of the difference was in the operating habits of the cars.

It appeared that humanity was upon the threshold of a new era, in which the value of personality would reappear among the things taken for granted. Strictly speaking, of course, Mahon machines were not persons. But they reflected the personalities of their owners. It might again seem desirable to be a decent human being if only because machines worked better for them.

But it would be tragic if Mahon machines were used to destroy humankind with themselves! Sergeant Bellews would have raged at the thought of training a Mahon unit to guide an atom bomb. It would have to be—in a fashion—deceived. He even disliked the necessity he faced that afternoon while a courier winged his way to the Pentagon with the top–secret tapes Betsy and Al and Gus had made.

The Rehab Shop was equipped not only to recondition machines but to test them. One item of equipment was a generator of substitute broadcast waves. It could deliver a carrier–wave down to half a micro–micro–watt of any form desired, and up to the power of a nearby transmitter. It was very useful for calibrating communicators. But Sergeant Bellews modified it to allow of variations in type as well as frequency and amplitude.

"I'm betting," he grunted, "that there's different sorts of the wave–type those guys want us to broadcast. Like there's a spectrum of visible light. If we were color–blind and yellow'd bust things, they'd transmit in red that we could see, and they'd tell us to broadcast something in yellow that'd wipe us out. And we wouldn't have sense enough not to broadcast the yellow, because we wouldn't know the difference between it and red until we did broadcast. Then it'd be too late."

Howell watched with tight–clamped jaws. He had committed himself to the authenticity of the broadcasts claiming to be from a future time. Now he was shaken, but only enough to admit the need for tests. Graves sat unnaturally still. Lecky looked at Sergeant Bellews with a peculiarly tranquil expression on his face.

"Only," grunted the sergeant, "it ain't frequency we got to figure, but type. Nobody hardly uses anything but sine waves for communication, but I got to make this gadget turn out a freak wave–type by guess and golly. I got a sort of test for it, though."

* * * * *

He straightened up and connected a cable from the generator to the Mark IV communicator which was a factory twin of Betsy.

"I'm gonna feed this communicator half a micro–micro–watt of stuff like the broadcast—I think," he announced grimly. "I saw the diagrams of the transmitters they want us to make. I'm guessing the broadcast–wave they use is close to it but not exact. Close, because it's bad for machines. Not exact, because they're alive while they use it. I hope I don't hit anything on the nose. Okay?"

Lecky said gently:

"I have never been more frightened. Go ahead!"

Sergeant Bellews depressed a stud. The communicator's screen lighted up instantly. It was receiving the generator's minute output and accepted it as a broadcast. But the signal was unmodulated, so there was no image nor any sound.

*****

The communicator's standby light flickered steadily.

Sergeant Bellews adjusted a knob on the generator. The communicator's standby flicker changed in amplitude. Bellews turned the knob back. He adjusted another control. The standby light wavered crazily.

Graves said nervously:

"I think I see. You are trying to make this communicator react as Betsy did. When it does, you will consider that your generator is creating a wave like the broadcasts from nowhere."

"Yeah," said Bellews. "It ain't scientific, but it's the best I can do."

He worked the generator–controls with infinite care. Once the communicator's standby light approached sine–wave modulation. He hastily shifted away from the settings which caused it. He muttered:

"Close!"

Then, suddenly, the communicator's lamp began to waver in an extraordinary, hysterical fashion. Sergeant

Bellews turned down the volume swiftly. He wiped sweat off his forehead.

"I—I think I got the trick," he said heavily. "It's a hell of a wave–type! Are you guys game to feed it into this communicator's output amplifier?"

"I have six sets of cold chills running up and down my spine," said Lecky. "I think you should proceed."

Howell said angrily:

"It's got to be tried, hasn't it?"

"It's got to be tried," acknowledged Sergeant Bellews.

He shifted the generator's cable from the communicator's input to the feed–in for preamplified signal. The communicator's screen went dark. It no longer received a simulated broadcast signal. It was now signalling—calling. But the instant the new signal started out, the standby light flickered horribly. Sergeant Bellews grimly plugged in other machines—to the three scientists they looked like duplicates of Gus and Al—to closed–circuit relationship with Betsy's twin. The standby light calmed.

"Now we test," he said grimly. "Got a watch?"

Lecky extended his wrist.

"Watch it," said Sergeant Bellews.

He stepped up the output.

"My watch has stopped," said Lecky, through white lips.

Graves looked at his own watch. He shook it and held it to his ear. He looked sick. Howell growled and looked at his own.

"That wave stops watches," he admitted unwillingly.

"But not Mahon machines easy," said Sergeant Bellews heavily, "and not us. There was almost three micro–micro–watts goin' out then. That's three–millionths of a millionth of a ampere–second at one volt. We—"

* * * * *

His voice stopped, as if with a click. The screen of Betsy's factory–twin communicator lighted up. A man's face peered out of it. He was bearded and they could not see his costume, but he was frightened.

"What—what is this?" cried his voice shrilly from the speakers.

Sergeant Bellews said very sharply:

"Hey! You ain't the guy we've been talking to!"

The screen went dark. Sergeant Bellews put his hand over the microphone opening. He turned fiercely upon the rest.

"Look!" he snapped. "We were broadcastin' their trick wave—the wave they used to talk to us! And they picked it up! But they weren't expectin' it! They were set to pick up the wave they told us to transmit! See? That guy'll come back. He's got to! So we got to play along! He'll want to find out if we got wise and won't broadcast ourselves to death! If he finds out we know what we're doin', they'll parachute a transmitter down on us before we can do it to them! Back me up! Get set!"

He removed his hand from the microphone.

"Callin' 2180!" he chattered urgently. "Calling the guy that just contacted us! Come in, 2180! You're not the guy we've been talking to, but come in! Come in, 2180!"

Howell said stridently:

"But if that's 2180, how'd we parachute—".

Lecky clapped a hand over his mouth with a fierceness surprising in so small a man. He whispered desperately into Howell's ear. Graves absurdly began to bite his nails, staring at the communicator-screen. Sergeant Bellews continued his calling, ever more urgently.

His voice echoed peculiarly in the Rehab Shop. It seemed suddenly a place of resonant echoes. All the waiting, repaired, or to-be-rehabilitated machines appeared to listen with interest while Sergeant Bellews called:

"Come in, 2180! We been trying to reach you for a coupla weeks! We got somebody else instead of you, and they been talkin' to us, and they say that they're 3020 instead of 2180, but we've got to contact you! They don't know anything about that germ that's gonna mutate and bump us off! It's ancient history to them. We got to reach you! Come in, 2180!"

The flickering yellow lights of the machines wavered as if all the quasi-living machines were listening absorbedly. The Rehab Shop was full of shadows. And Sergeant Bellews sat before the dark-screened communicator with sweat on his face, calling cajolingly to nothingness to come in.

After five minutes the screen grew abruptly bright again. The brisk, raceless broadcaster of the earlier broadcast—not the bearded man—came back. He forced a smile:

"Ah! 1972! At last you reach us! But we did not hope you could make your transmitters so soon!"

"We tried to analyze your wave," said Sergeant Bellews, with every appearance of feverish relief, "but we only got it approximate. We tried callin' back with what we got, and we got through time, all right, but we contacted some guys in 3020 instead of you! We need to talk to you!—Can you give me the stuff about that bug that's gonna wipe out half of us? Quick? I got a recorder goin'."

* * * * *

The completely uncharacterizable man in the screen forced a second smile. He held something to his ear. It would be a tiny sound–receiver. Obviously the contact in time or place or nowhere was being viewed by others than the one man who appeared. He was receiving instructions.

"Ah!" he said brightly, "but now that you have the contact, you will not lose it again! Leave your controls where they are, and our learned men will tell your learned men all that they need to know. But—3020? You contacted 3020? That is not in our records of your time!"

He listened again to the thing at his ear. His expression

became suddenly suspicious, as if someone had ordered that as well as the words which came next.

"We do not understand how you could contact a time a thousand years beyond us. It is possible that you attempt a joke. A—a kid, as you would say."

*****

Sergeant Bellews beamed into the screen which so remarkably functioned as a transmitting-eye also.

"Hell!" he said cordially. "You know we wouldn't kid you! You or our great-great-great-grandchildren! We depend on you! We got to get you to tell us how not to get wiped out! In 3020 the whole business is forgotten. It's a thousand years old, to them! But they're passin' back some swell machinery—"

He turned his head as if listening to something the microphone could not pick up. But he looked appealingly at Lecky. Lecky nodded and moved toward the communicator.

"Look!" said Sergeant Bellews into the screen. "Here's Doc Lecky—one of our top guys. You talk to him."

He gave his seat to Lecky. Out of range of the communicator, he mopped his face. His shirt was soaked through by the sweat produced by the stress of the past few minutes. He shivered violently, and then clamped his teeth and fumbled out sheets of paper. He beckoned to Graves. Graves came.

"We—we got to give him a doctored circuit," whispered Sergeant Bellews desperately, "and it's got to be good—an' quick!"

Graves bent over the paper on which the sergeant dripped sweat. The sergeant murmured through now-chattering teeth what had to be devised, and at once. It must be the circuit–diagram for a transmitter to be given to the man whose face filled the screen. The transmitter must be of at least twenty–kilowatt power. It must be such a circuit as nobody had ever seen before.

It must be convincing. It should appear to radiate impossibly, or to destroy energy without radiation. But it must actually produce a broadcast signal of this exotic type—here the sergeant described with shaky precision the exact constants of the wave to be generated—and the broadcaster from nowhere must not be able to deduce those constants or that wave–type from the diagram until he had built the transmitter and tried it.

"I know it can't be done!" said the sergeant desperately. "I know it can't! But it's gotta be! Or they'll parachute a transmitter down on us sure."

Graves smiled a quick and nervous smile. He began to sketch a circuit. It was a wonderful thing. It was the product of much ingenuity and meditation. It had been devised—by himself—as a brain-teaser for the amusement of other high-level scientific brains. Mathematicians zestfully contrive problems to stump each other. Specialists in the higher branches of electronics sometimes present each other with diagrammed circuits which pretend to achieve the impossible. The problem is to find the hidden flaw.

Graves deftly outlined his circuit and began to fill in the details. Ostensibly, it was a circuit which consumed energy and produced nothing—not even heat. In a sense it was the exact opposite of a perpetual-motion scheme, which pretends to get energy from nowhere. This circuit pretended to radiate energy to nowhere, and yet to get rid of it.

*****

Presently Lecky could be heard expostulating gently:

"But of course we are willing to give you the circuit by which we communicate with the year 3020! Naturally! But it seems strange that you suspect us! After all, if you do not tell us how to meet the danger your broadcasts have told of, you will never be born!"

Sergeant Bellews mopped his face and moved into the screen's field of vision.

"Doc," he said, laying a hand on Lecky's arm. "Doc Graves is sketchin' what they want right now. You want to come show it, Doc?"

Graves took Lecky's place. He spread out the diagram, finishing it as he talked. His nervous, faint smile appeared as the mannerism of embarrassment it was.

"There can be no radiation from a coil shaped like this," he said embarrassedly, "because of the Werner Principle…. Yet on examination …input to the transistor series involves … energy must flow … and when this coil…."

His voice flowed on. He explained a puzzle, presenting it diffidently as he had presented it to other men in his own field. Then he had been playing—for fun. Now he played for perhaps the highest stakes that could be imagined.

He completed his diagram and, smiling nervously, held it up to the communicator–screen. It was instantly transmitted, of course. To nowhere. Which was most appropriate, because it pretended to be the diagram of a circuit sending radiation to the same place.

* * * * *

The face on the screen twitched, now. The hand with the tiny earphone was always at the ear of the man on the screen, so that he plainly did not speak one word without high authority.

"We will—examine this," he said. His voice was a full two tones higher than it had been. "If you have been—truthful we will give you the information you wish."

Click! The screen went dark. Lecky let out his breath. Sergeant Bellews threw off the transmission switch. He began to shake. Howell said indignantly:

"When I make a mistake, I admit it! That broadcast isn't from the future! If it hadn't been a lie, he'd have known he had to tell us what we wanted to know! He couldn't hold us up for terms! If he let us die he wouldn't exist!"

"Y–yeah," said Sergeant Bellews. "What I'm wonderin' is, did we fool him?"

"Oh, yes!" said Graves, with diffident confidence. "I don't know but three men in the world who could find the flaw in that circuit." He smiled faintly. "But it radiates all the energy that's fed into it." He turned to

Sergeant Bellews. "You gave me the constants of a wave you wanted it to radiate. I fixed it. It will. But why that special type—that special wave?"

Sergeant Bellews pulled himself together.

"Because," he said grimly, "that was the wave they wanted us to broadcast. What I'm hoping is that you gave 'em a transmitter to do exactly the same thing as the one they designed for us. If they're fooled, they'll broadcast the wave they told us to broadcast. If it busts machines, it'll bust their machines. If it stops all dynamic systems dead—includin' men—they'll be stopped dead, too." Then he looked from one to another of the three scientists, each one reacting in his own special way. "Personally," said Sergeant Bellews doggedly, "I'm goin' to have a can of beer. Who'll join me?"

* * * * *

The world wagged on. The automatic monitors in Communications Center reported that another broadcast had been received by Betsy and undoubtedly unscrambled by Al and Gus, working as a team. The reported broadcast was, of course, an interception of the two–way talk from the Rehab Shop.

The tall young lieutenant, working with his eyes kept

conscientiously shut, extracted the tapes and loaded them in a top-security briefcase. A second courier took off for Washington with them. There a certified, properly cleared major-general had them run off, and saw and heard every word of the conversation between the Rehab Shop and—nowhere. He howled with wrath.

Sergeant Bellews went into the guardhouse while plane-loads of interrogating officers flew from Washington. Howell and Graves and Lecky went under strict guard until they could be asked some thousands of variations of the question, "Why did you do it?" The high brass quivered with fury. They did not accept decisions made at non-commissioned-officer level.

Communication with their great-great-great-grandchildren, they considered, should have been begun with proper authority and under high-ranking auspices. They commanded that 2180 should immediately be re-contacted and properly authorized and good-faith conference begun all over again. The only trouble was that they could get no reply.

The dither was terrific and the tumult frantic. When, moreover, even Betsy remained silent, and Al and Gus had nothing to unscramble, the high brass built up explosive indignation. But it was confined to top-security levels.

The world outside the Pentagon knew nothing. Even at Research Installation 83 very, very few persons had the least idea what had taken place. The sun shone blandly upon manicured lawns, and the officers' children played vociferously, and washing-machines laundered diapers with beautiful efficiency, and vacuum cleaners and Mahon-modified jeeps performed their functions with an air of enthusiastic contentment. It seemed that a golden age approached.

It did. There were machines which were not merely possessions. Mahon-modified machines acquired reflections of the habits of the families which used them. An electric icebox acted as if it took an interest in its work. A vacuum cleaner seemed uncomfortable if it did not perform its task to perfection. It would seem as absurd to exchange an old, habituated family convenience as to exchange a member of the family itself. Presently there would be washing-machines cherished for their seeming knowledge of family-member individual preferences, and personal fliers respected for their conscientiousness, and one would relievedly allow an adolescent to drive a car if it were one of proven experience and sagacity....

*****

The life of an ordinary person would be enormously enriched. A Mahon-modified machine would not even wear out. It took care of its own lubrication and upkeep—giving notice of its needs by the behavior of its standby–lamp. When parts needed replacement one would feel concern rather than irritation. There would be a personal relationship with the machines which so faithfully reflected one's personality.

And the machines would always, always, always act toward humans according to the golden rule.

But meanwhile the Rehab Shop was taken over by officers of rank. They tried frantically to resume the communication that had been broken off. Suspecting that Sergeant Bellews had shifted controls, they essayed to shift them back. The communicator which was Betsy's factory twin went into sine–wave standby–modulation, and suddenly smoked all over and was wrecked. The wave–generator went into hysterics and produced nothing whatever. Then there was nothing to do but pull Sergeant Bellews out of the clink and order him to do the whole business all over again.

"I can't," said Sergeant Bellews indignantly. "It can't be done. Those guys are busy buildin' a transmitter according to the diagram Doc Graves gave them. They won't pay no attention to anything until they'd tried to chat with their great–great–great–grand–children in 3020. They were phonys, anyhow! Pretendin' to be in 2180 and not knowin' what Mahon units could do!"

Lecky and Graves and Howell were even less satisfactory. They couldn't pretend even to try what the questioning–teams from the Pentagon wanted them to do. And Betsy remained silent, receiving nothing, and Gus and Al waited meditatively for something to unscramble, and nothing turned up.

And then, at 3:00 P.M. Greenwich mean time, on August 9, 1972, nearly every operating communicator in the fringe of free nations around the territory of the Union of Communist Republics—all communicators blew out.

There were only four men in the world who really knew why—Sergeant Bellews and Lecky and Graves and Howell. They knew that somewhere behind the Iron Curtain a twenty–kilowatt transmitter had been turned on. It produced a wave of the type and with the characteristics that would have been produced by a transmitter built from the diagram sent through Betsy and Al and Gus for people in the United States to build. Obviously, it had been built from Graves' diagram broadcast to somewhere else and it broadcast what the United States had been urged to broadcast.

*****

It blew itself out instantly, of course. The wave it produced would stop any dynamic system at once, including its own. But it hit Stockholm and traffic jammed as the dynamic systems of cars in operation were destroyed. In Gibraltar, the signal–systems of the Rock went dead. All around the fringe of the armed Communist republics machines stopped and communications ended and very many persons with heart conditions died very quietly. Because their dynamic systems were least stable. But healthy people—like Mahon–modified machines—had great resistance … outside the Iron Curtain.

There was, though, almost a vacuum of news and mechanical operations at the rim of a nearly perfect circle some four thousand miles in diameter, whose center was in a Compub research installation.

It was very bad. Such a panic as had never been known before swept the free world. Some mysterious weapon, it was felt, had been used to cripple those who would resist invasion, and the Compub armed forces would shortly be on the march, and Armageddon was at hand. The free world prepared to die fighting.

But war did not come. Nothing happened at all. In three days there were sketchy communications almost everywhere outside that monstrous circle of silence. But nothing came out of that circle. Nothing.

In two weeks, exploring parties cautiously crossed the barbed–wire frontier fences to find out what had happened. Those who went farthest came back shaken and sick. There were survivors in the Compubs, of course. Especially near the fringes of the circle. There were some millions of survivors. But there was no longer a nation to be called the Union of Communist Republics. There were only frightened, starving people trudging blindly away from cities that were charnel-houses and machines that would not run and trees and crops and grasses that were stark dead where they stood. It would be a long time before anybody would want to cross those lifeless plains and enter the places which once had been swarming hives of homes and people.

* * * * *

And presently, of course, Sergeant Bellews was let out of the guardhouse. He could not be charged with any crime. Nor could Graves nor Lecky nor Howell. They were asked, confidentially, to keep their mouths shut. Which they would have done anyhow. And Sergeant Bellews was asked with reluctant respectfulness, just what he thought had really happened.

"Some guys got too smart," he said, fuming. "A guy that'll broadcast a wave that'll wreck machines ... I haven't got any kinda use for him! Dammit, when a machine treats you accordin' to the golden rule, you oughta treat it the same way!"

There were other, also-respectful questions.

"How the hell would I know?" demanded Sergeant Bellews wrathfully. "It coulda been that we did make contact with 2180, and they were smart an' told the Compubs to try out what we told 'em. But I don't believe it. It coulda been a kinda monster from some other planet wanting us wiped out. But he learned him a lesson, if he did! And o' course, it coulda been the Compubs themselves, trying to fool us into committing suicide so they'd—uh—inherit the earth. I wouldn't know! But I bet there ain't any more broadcasts from nowhere!"

He was allowed to return to the Rehab Shop, and the flickering standby lights of many Mahon-modified machines seemed to glow more warmly as he moved among them.

And he was right about there not being any more broadcasts from nowhere.

There weren't.

Not ever.

THE END